永不懈怠的玩樂之心

　　舊曆年前，大姊抱來一團毛絨絨、幾分像鼠、幾分像貓、幾分像浣熊、又有幾分肖似眼鏡猴的小傢伙，一時之間說不準那究竟是什麼。全家人多年來難得有個共同話題，這時紛紛發表揣測，最後做出他是隻狗的結論。不過有些家庭成員不大服氣，而且接連幾天這隻被推斷為狗的小傢伙都沒吭半聲，我們頗擔心他突然「喵」出來，全然推翻先前的結論，於是幾個已經有些年紀的歐吉桑與歐巴桑不時衝著他汪汪叫，企圖說服他：「你真的是隻狗」，以保存自己久經風霜的顏面。

　　「可樂」是他的名字，那時他還不滿兩個月大，渾不怕生，與我更是投緣，初見面，便對我跟前繞後，舅甥倆一拍即合，如膠似漆，其養父母竟因此有些兒吃味哩！其實他當時已生命垂危，腸子長了蟲子，非但腹瀉不止，還幾乎飲食不進。然而即使年稚病弱，在醒著的每一刻，他還是不停地玩耍，熱切地探索這個新奇的世界，並且帶給我們無限的歡樂。

　　大姊察覺他的狀況日益惡化，不管仍在新年假期當中，哪顧淒風苦雨、春寒料峭，噙著淚，懷抱他滿街亂走，尋找還開著的獸醫院。終於覓著一家，不過醫生檢查後說沒把握救得了，得住院觀察幾天。另一個外甥（用兩條腿走路、身高一米八的那位）放學回家沒見到可樂，忙問他娘，他娘抽抽噎噎地說了，外甥臉色突變，立即奔到廁所裡放聲大哭。

　　當然，可樂終究無恙，而今才有這本小書。書裡「人間的可樂狗」一詩雖是脫胎自「人間的四月天」，然而我實無意將可樂與徐志摩或梁從誡相提並論（咸信「四月天」詩乃為其中之一而作），畢竟可樂比這兩位先生都年輕可愛許多。徐是赫赫有名的大詩人，那就甭提了。梁教授為聞人之後，梁啟超之嫡孫，還有個才貌兼具的母親林徽音。但我家可樂的先祖乃「維多利亞女王之愛犬」，他並備有血統證明，論起家世，也不遑稍讓哩！況且可樂還有機會參加選美，這碼子勾當，徐梁二公可就望塵莫及了。

　　我還撿拾了數十則西人喻狗名言，其味堪稱雋永，又不辭才薄，譯成中文。有句

話說：「翻譯就像女人，愈漂亮的愈不忠實（男人的忠實度則與相貌無關）。」這當然是譯筆粗疏的絕佳託辭。所附相片皆生活上的浮光掠影，非刻意拍攝，並無出奇之處，因此借用登山名家高銘和先生的幾幅大作來撐撐場面。老狗變不出新把戲，我玩起合成圖片的新把戲確實不怎麼得心應手，不過誰在乎呢，好玩就好！其實就是可樂那顆永不懈怠的玩樂之心，讓我漸趨枯乏的生活又綻放出光輝，至少搞這樣一本書能消耗些心神，打發些時間，跟他玩耍更是樂趣無窮。

　　也許有人會問，狗兒除了吃喝拉撒睡，會做的不外遊戲與交配（真是神仙般生活），生命不知有何意義？班‧威廉斯（Ben Williams）曾說：「世上沒有任何精神科醫師比得上舔你臉的小狗（There is no psychiatrist in the world like a puppy licking your face）。」所以哪天若有狗兒出來掛牌執業，大家犯不著大驚小怪。不過要是精神科醫師也想學狗兒舔您尊臉，可別吝於賞他一巴掌，順便罵聲："You, bad dog!"

人間的可樂狗

我說你是人間的可樂狗，
淘氣，惹笑了四面佛。
輕靈，在我的
懷抱中不停地扭。

你是可樂瓶裡的咖啡因，
一團毛絨絨的球，
大便隨意地洩漏，
小便揮灑在四周。

犯錯，闖了禍，
你呀，腳底趕緊抹滑油。

4

要打你，著實不忍，
罵你，好似彈琴對頭牛。

這熱麻癢的輕舐，消愁；
那晃擺搖的尾巴，解憂。
柔嫩喜悅，彷彿
搖曳在心湖裡的一葉舟。

你是頑皮精靈的嬉戲，
慧黠仙子的凝眸──
你是愛，是暖，是守候，
你是人間的可樂狗。

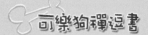
可樂狗禪逗畫

Dog Quotes

狗語錄

The one absolutely unselfish friend that man can have in this selfish world, the one that never deserts him, the one that never proves ungrateful or treacherous, is his dog.

~~ *George Graham Vest*

身處於如此自私自利的世界中，一個人所能擁有的最為無私、永遠不離不棄、絕不忘恩負義、一定不會背叛的朋友，就是他的狗。

~~喬治・葛拉罕・衛斯特
十九世紀美國參議員

Dogs are our link to paradise. They don't know evil or jealousy or discontent. To sit with a dog on a hillside on a glorious afternoon is to be back in Eden, where doing nothing was not boring — it was peace.

~~ *Milan Kundera*

不懂得邪惡、嫉妒或忿懣不滿的狗兒，是我們通往天堂樂土的媒介。明媚怡人的午後，與愛犬同坐在山坡上，就彷彿回到伊甸園一般，在那兒，無所事事並不枯燥乏味，而是讓人感受到恬適平和。

～～米蘭‧昆德拉
捷克當代知名作家

If you are a dog and your owner suggests that you wear a sweater ... suggests that he wear a tail.

~~ *Fran Lebowitz*

狗呀！你的主人要你穿上衣服的話，就叫他也長條尾巴看看。

~~芙蘭‧李柏葳姿
美國當代幽默作家

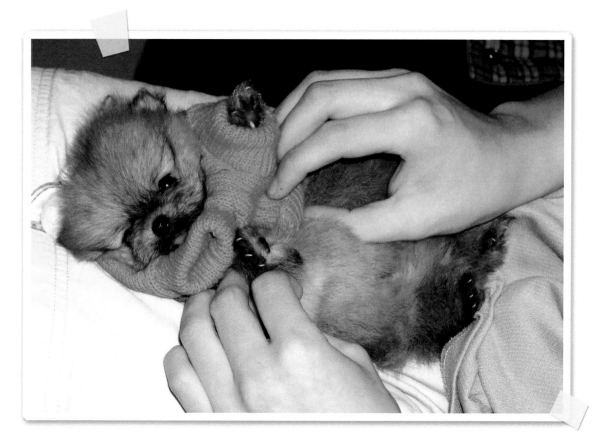

The more I see of men the more I like dogs.

~~ *Madame de Staël*

男人看愈多，我就愈喜歡狗。

~~ 思達爾夫人
十九世紀法國作家

My dog can bark like a congressman, fetch like an aide, beg like a press secretary and play dead like a receptionist when the phone rings.

~~ *Gerald B. H. Soloman, U.S. Congressman*

我的狗能叫囂如立委，跑腿似助理，求懇像新聞祕書，還會裝出死樣子，就好比電話鈴響時的櫃檯人員。

～～傑羅‧索羅門
美國國會議員

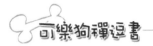
可樂狗禪逗書

You think dogs will not be in heaven? I tell you, they will be there long before any of us.

~~ *Robert Louis Stevenson*

你以為狗上不了天堂？告訴你，我們都還沒上去前，他們早就在那裡啦！

~~羅勃・路易士・史蒂文生
《金銀島》作者

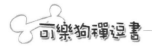

No matter how little money and how few professions you own, having a dog makes you rich.

~~ *Louis Sabin*

就算你再窮、再沒本事，擁有隻狗便讓你富足。

~~路易士‧沙賓
美國當代作家

可樂狗禪逗書

Yesterday I was a dog. Today I'm a dog. Tomorrow I'll probably still be a dog. Sigh! There is so little hope for advancement.

~~ Snoopy

昨日乃狗，今日為狗，明日亦狗。噫！進步之望幾希。

~~史努比
世界最著名的小獵犬

Properly trained, a man can be dog's best friend.

~~ Corey Ford

善加訓練，人也可以成為狗的最好朋友。

~~柯利・福特
美國當代作家

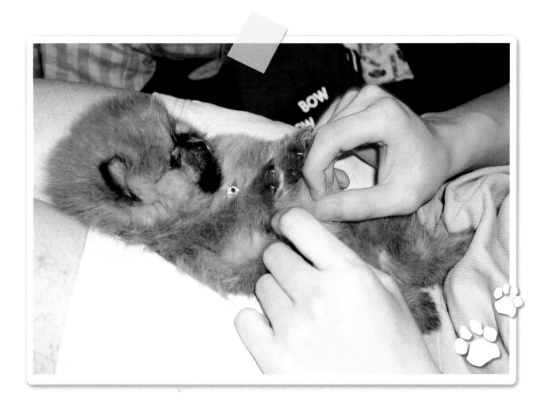

If dogs could talk it would take a lot of the fun out of owning one.

~~ *Andy Rooney*

要是狗會說人話，養狗也許就沒那麼好玩了。

~~安迪‧魯尼
美國資深媒體人

Acquiring a dog may be the only opportunity a human ever has to choose a relative.

~~ *Mordecai Siegal*

養狗可能是人類能夠選擇親戚的唯一機會。

~~ 摩得開‧席格
美國寫犬人協會理事長

In order to really enjoy a dog, one doesn't merely try to train him to be semi-human. The point of it is to open oneself to the possibility of becoming partly a dog.

~~ *Edward Hoagland*

要真正享受有狗為伴的樂趣，不能僅是將他訓練得人模人樣，重點是要敞開胸懷，讓自己也有機會變成狗模狗樣。

~~愛德華・侯格蘭
美國當代作家

The old saw about old dogs and new tricks only applies to certain people.

~~ Daniel Pinkwater

「老狗變不出新把戲」這句老諺語,只適用於某些人。

~~丹尼爾‧頻克瓦特
童書作家

Dogs read the world through their noses and write their history in urine.

~~ *J. R. Ackerley, My Dog Tulip*

狗以鼻子閱讀此世界，用尿液撰寫其歷史。

~~ 艾克里
於My Dog Tulip 一書

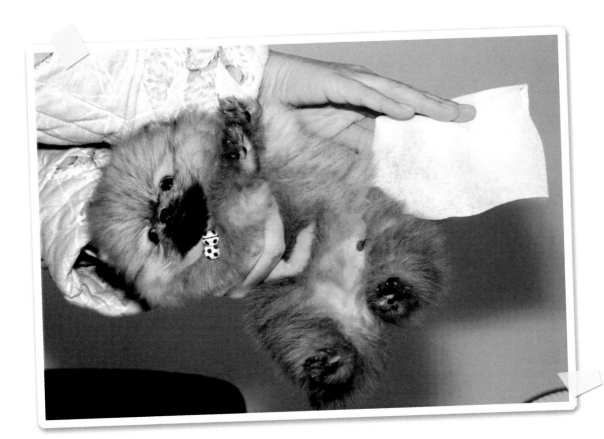

I can train any dog in 5 minutes. It's training the owner that takes longer.

~~ *Barbara Woodhouse*

教狗容易，教飼主難。

＼＼ 芭芭拉‧伍德豪斯
作者‧犬科專家

My husband and I are either going to buy a dog or have a child. We can't decide whether to ruin our carpets or ruin our lives.

~~ *Rita Rudner*

我們夫妻倆想要買條狗，或者生個孩子，但我們無法決定到底是要犧牲地毯，還是犧牲生活。

~~麗塔‧魯德那
英國女諧星

Do not make the mistake of treating your dogs like humans or they will treat you like dogs.

~~ *Martha Scott*

別錯把狗當人對待，否則他們會把你當狗對待。

~~瑪莎‧史考特
美國老牌女演員

The dog who meets with a good master is the happier of the two.

~~ *Maeterlinck*

狗逢良主，相得益彰。

~~梅特林克
諾貝爾文學獎得主

The dog has got more fun out of man than man has got out of the dog, for man is the more laughable of the two animals.

~~ James Thurber

狗從人身上得到的樂趣多於人從狗身上得到的樂趣，因為人是這兩者中比較可笑的一種。

~~詹姆士‧瑟伯爾
政治學教授

可樂狗禪逗書

Don't accept your dog's admiration as conclusive evidence that you are wonderful.

~~ *Ann Landers*

別因為你的狗對你敬仰有加就自命不凡。

~~安‧蘭德斯
女專欄作家

可樂狗禪逗書

The reason a dog has so many friends, is that he wags his tail instead of his tongue.

~~*Anonymous*

狗能廣結善緣，因為他勤於搖擺的是尾巴而非舌頭。

~~佚名

A dog can express more with his tail in minutes than his owner can express with his tongue in hours.

~~ Anonymous

狗搖一搖尾，勝於飼主的萬語千言。

~~佚名

One reason a dog is such comfort when you're downcast is that he doesn't ask to know why.

~~Anonymous

當你垂頭喪氣時，狗是如此好慰藉的原因之一是，他不會問你怎麼了。

~~佚名

Dogs have given us their absolute all. We are the center of their universe. We are the focus of their love and faith and trust. They serve us in return for scraps. It is without a doubt the best deal man has ever made.

~~ *Roger Caras*

狗對我們毫無保留，我們是他們的世界中心，其信望愛之所鍾。僅用殘羹剩飯便換得他們的全心服侍，毋庸置疑，這是人類做過的最划算的一椿買賣。

~~羅傑‧卡拉斯
電影工作者，作家

Every boy who has a dog should also have a mother, so the dog can be fed regularly.

~~ *Anonymous*

每個有狗的小孩也該有娘，因此狗才不會有一餐沒一餐。

~~佚名

A piece of grass a day keeps the vet away.

~~ *Unknown Dog*

每天啃根草，獸醫不見了 。

～～無名犬

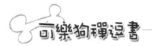
THINGS WE CAN LEARN FROM A DOG

以狗為師

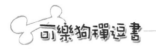

Eat with gusto and enthusiasm.

吃得津津有味。

When loved ones come home, always run to greet them.

所愛之人一到家，總會跑去歡迎。

When it's in your best interest, practice obedience.

當順從於己有利時，就識相點兒。

Take naps and stretch before rising.

小睡一會兒，起身前先伸伸懶腰。

Let others know when they've invaded your territory.

當有人侵犯你的領域時，讓他們得到警告。

Be loyal.

忠誠。

Run, romp and play daily.

每天跑一跑，跳一跳，玩一玩，鬧一鬧。

Never pretend to be something you're not.

從不裝模作樣。

If what you want lies buried, dig until you find it.

鍥而不捨。

When someone is having a bad day, be silent, sit close by and nuzzle him gently.

當某人情緒不佳時，閉上狗嘴，依偎在他身旁，輕柔地撫慰他。

Avoid biting when a simple growl will do.

動口就足夠的話，犯不著動手。

On hot days, drink lots of water and lay under a shady tree.

天熱時，喝大量水，躺在林蔭下。

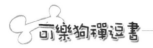
可樂狗禪逗書

When you're happy, dance around and wag your entire body.

高興時跳跳舞，搖擺整個身軀。

No matter how often you're scolded, don't buy into the guilt thing and pout...
run right back and make friends.

不管多常挨罵，切莫自責，而且馬上重修舊好。

Delight in the simple joy of a long walk.

享受漫步的單純樂趣。

語言文學類　PG0061

可樂狗禪逗書

作　　者 / 傅達德
發 行 人 / 宋政坤
責任編輯 / 莊芯媚
美術編輯 / 莊芯媚
數位轉譯 / 徐真玉、沈裕閔
圖書銷售 / 林怡君
法律顧問 / 毛國樑 律師

出版印製 / 秀威資訊科技股份有限公司
　　　　　台北市內湖區瑞光路583巷25號1樓
　　　　　電話：02-2657-9211　傳真：02-2657-9106
　　　　　E-mail：service@showwe.com.tw
經 銷 商 / 紅螞蟻圖書有限公司
　　　　　台北市內湖區舊宗路二段121巷28、32號4樓
　　　　　電話：02-2795-3656　傳真：02-2795-4100
　　　　　http://www.e-redant.com

2005 年 5 月　BOD 一版
定價：150元

國家圖書館出版品預行編目

可樂狗禪逗書 / 傅達德作.　--　一版.　--　臺北
　市 ： 秀威資訊科技，　2005[民94]
　　　面 ；　　公分.　--　（語言文學 ； PG0061）

　ISBN 978-986-7263-38-4（平裝）

　1.　犬 - 通俗作品

437.66　　　　　　　　　　　　94009040